SPACE FRONTIERS

Our Solar System

Helen Whittaker

A+

This edition first published in 2011 in the United States of America by Smart Apple Media.

All rights reserved. No part of this book may be reproduced in any form or by any means without written permission from the Publisher.

Smart Apple Media
P.O. Box 3263
Mankato, MN, 56002

First published in 2010 by
MACMILLAN EDUCATION AUSTRALIA PTY LTD
15-19 Claremont Street, South Yarra 3141

Visit our website at www.macmillan.com.au or go directly to www.macmillanlibrary.com.au

Associated companies and representatives throughout the world.

Copyright © Macmillan Publishers Australia 2010

Library of Congress Cataloging-in-Publication Data

Whittaker, Helen.
    Our solar system / Helen Whittaker.
        p. cm. — (Space frontiers)
    Includes index.
    ISBN 978-1-59920-572-4 (library binding)
    1. Solar system—Juvenile literature. I. Title.
    QB501.3.W46 2010
    523.2—dc22
                                2009038478

Edited by Laura Jeanne Gobal
Text and cover design by Cristina Neri, Canary Graphic Design
Page layout by Cristina Neri, Canary Graphic Design
Photo research by Brendan and Debbie Gallagher
Illustrations by Alan Laver, except page 16 by Melissa Webb

Manufactured in China by Macmillan Production (Asia) Ltd.
Kwun Tong, Kowloon, Hong Kong
Supplier Code: CP December 2009

Acknowledgments
The author and the publisher are grateful to the following for permission to reproduce copyright material:

Front cover photos of the planets in our solar system © Brandon Alms/iStockphoto; blue nebula background © sololos/iStockphoto.

Photographs courtesy of:
ESA/NASA/JPL/University of Arizona, **23**; © Brandon Alms/iStockphoto, **3**; NASA, ESA and G. Bacon (STScI), **28**; NASA, ESA, and A. Feild (STScI), **24**; NASA/HSF, **17**; NASA/Johns Hopkins University Applied Physics Laboratory/Arizona State University/Carnegie Institution of Washington, **11**; NASA/JPL, **10**, **16** (bottom), **20**, **26**, **27**; NASA/JPL-Caltech, **25**; NASA/JPL-Caltech/University of Arizona/Texas A&M University, **18**; NASA/JPL/DLR, **21**; NASA/JPL/Space Science Institute, **22** (left); NASA/JSC, Eugene Cernan, **4**; Steve Lee (University of Colorado), Jim Bell (Cornell University), Mike Wolff (Space Science Institute) and NASA, **19**; NASA/Lunar and Planetary Laboratory, **6**; Wolcott Henry/National Geographic Stock, **15**; Photodisc, **13** (right); Photolibrary/Photo Researchers, **5**, back cover; Photolibrary/Jerry Schad, **12**; Photolibrary/NASA/SPL, **13** (left); Photolibrary/Richard J Wainscoat, **29**; SOHO (ESA & NASA), **9**; Spitzer Space Telescope, **30**.

Images used in design and background on each page © prokhorov/iStockphoto, Soubrette/iStockphoto

While every care has been taken to trace and acknowledge copyright, the publisher tenders their apologies for any accidental infringement where copyright has proved untraceable. Where the attempt has been unsuccessful, the publisher welcomes information that would redress the situation.

Contents

Space Frontiers   4
Our Solar System   5
Mapping the Solar System   6

The Sun   8
Mercury   10
Venus   12
Earth   14
The Moon   16
Mars   18
Jupiter   20
Saturn   22
Uranus   24
Neptune   26
Other Bodies   28

The Future of Our Solar System   30
Glossary   31
Index   32

Glossary Words

When a word is printed in **bold**, you can look up its meaning in the Glossary on page 31.

Space Frontiers

A frontier is an area that is only just starting to be discovered. Humans have now explored almost the entire planet, so there are very few frontiers left on Earth. However, there is another frontier for us to explore and it is bigger than we can possibly imagine—space.

Where Is Space?

Space begins where Earth's **atmosphere** ends. The atmosphere thins out gradually, so there is no clear boundary marking where space begins. However, most scientists define space as beginning at an altitude of 62 miles (100 km). Space extends to the very edge of the universe. Scientists do not know where the universe ends, so no one knows how big space is.

Exploring Space

Humans began exploring space just by looking at the night sky. The invention of the telescope in the 1600s and improvements in its design have allowed us to see more of the universe. Since the 1950s, there has been another way to explore space—spaceflight. Through spaceflight, humans have **orbited** Earth, visited the Moon, and sent space probes, or small unmanned spacecraft, to explore our **solar system**.

▲ Spaceflight is one way of exploring the frontier of space. Astronaut Harrison Schmitt collects Moon rocks during the Apollo 17 mission in December 1972.

Our Solar System

The solar system is our neighborhood in space. At its center is a star, the Sun. Orbiting the Sun are eight planets and their **moons**, as well as several **dwarf planets** and millions of **asteroids** and **comets**.

How Did the Solar System Form?

The solar system formed about 4.6 billion years ago from the solar nebula, a huge disc of gas and dust left behind when the Sun was formed. Near the Sun, where **gravity** was stronger, heavier **elements** gathered together to form the terrestrial, or rocky, inner planets (Mercury, Venus, Earth, and Mars). Farther out, where gravity was weaker, lighter elements formed the **gas giants** (Jupiter, Saturn, Uranus, and Neptune).

Did You Know?

Our solar system is a part of the Milky Way **galaxy**, which is just one of more than 100 billion galaxies in the universe.

How Big Is the Solar System?

The solar system is incredibly big. Scientists are still discovering objects farther and farther away, which makes it difficult to confirm exactly how big our solar system is. Sedna, currently the most distant object in the solar system, is about 88 times farther away from the Sun than Earth is. A car traveling at 62 miles (100 km) per hour would take more than 15,000 years to get there!

▼ This photograph shows four members of our solar system—Venus, Mars, Jupiter, and the Moon—in the sky over Arizona.

MAPPING OUR SOLAR SYSTEM

It is difficult to show the Sun and all the planets drawn to the same scale, illustrating their relative sizes and the distances between them accurately. This is because the distances between them are so vast.

Solving the Problem of Distance and Size

If Mercury, the planet closest to the Sun, were drawn with a diameter of just 0.08 inches (2 mm), the Sun would have to be drawn with a diameter of 22 inches (58 cm) and Mercury would have to be placed more than 79 feet (24 m) away from it!

This problem can be solved by drawing two diagrams, one that shows the relative sizes of the planets and another that shows the relative distances between their orbits.

▼ This diagram shows the approximate relative sizes of the Sun and the planets. The distances between them are not to scale.

▼ **This diagram shows the approximate relative distances between the orbits of the planets. The sizes of the Sun and planets are not to scale.**

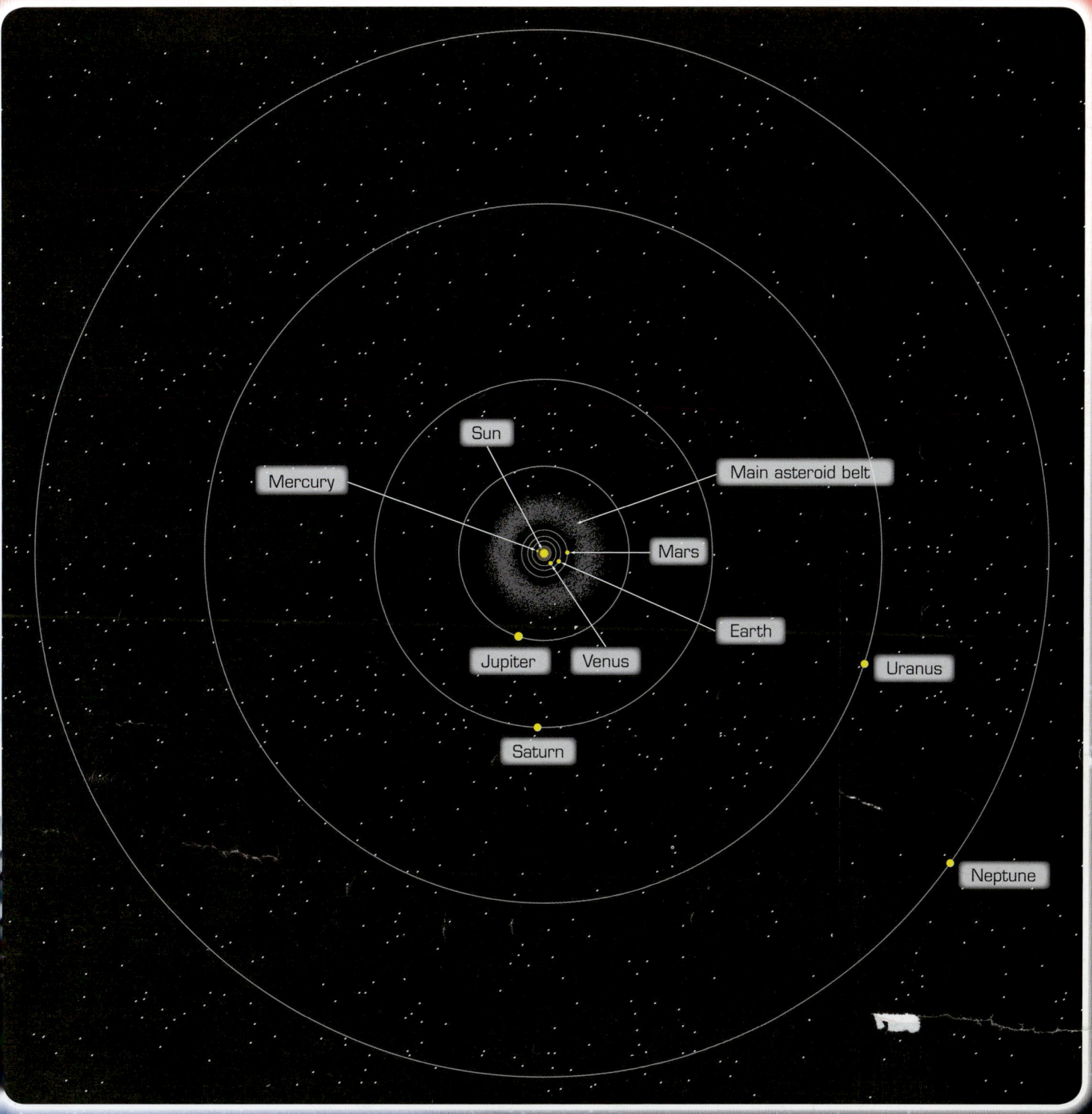

THE SUN

The Sun is a star. It is just one of more than 200 billion stars that make up the Milky Way galaxy. Without the Sun, there would be no life on Earth. In fact, there would be no Earth at all!

Why Does the Sun Shine?

The Sun is a huge ball of gas. Its mass is so large that the pressure and heat at its core are high enough to start a process called **nuclear fusion**. It is this process that causes the Sun to release vast amounts of energy in the form of heat, light, and other types of **electromagnetic radiation**. Stars that produce energy through nuclear fusion at their core are called main sequence stars.

The Sun's Life Cycle

The Sun is roughly 4.6 billion years old and is about halfway through its life as a main sequence star. In 5 billion years, it will enter the next phase of its existence, swelling up to become a **red giant**. Eventually it will start to shed its outer layer and become a planetary nebula, which is the burned-out core surrounded by a glowing shell of gas. Finally, it will collapse to become a **white dwarf** and will slowly fade.

Did You Know?

Looking directly at the Sun could permanently damage a person's eyes.

▼ This diagram shows the life cycle of our Sun, which is gradually getting warmer. Scientists believe that in about 500 million years, Earth will be too hot to support life.

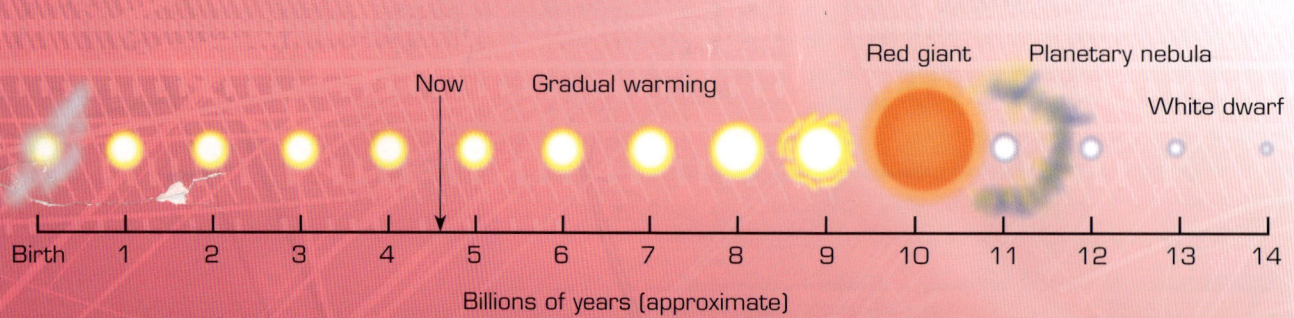

▲ The Sun's surface has sunspots and active regions. Sunspots are cool, dark areas, while active regions are hot, light areas.

Features of the Sun

The Sun's surface features are always changing, with sunspots (cool, dark areas) and active regions (hot, light areas) constantly on the move. The Sun experiences explosive storms which fling enormous plumes of superheated and electrically charged gas, known as plasma, into space. Localized storms are called solar flares and widespread events are known as coronal mass ejections (CMEs).

Facts and Figures

Name: the Sun

Type of body: star

Age: 4.6 billion years

Diameter: 864,327 mi (1.39 million km)

Volume: 1.3 million Earths could fit inside

Composition: 92.1% hydrogen, 7.8% helium

Average distance from Earth: 92.96 million mi (149.6 million km)

Average core temperature: more than 27 million degrees F (15 million degrees C)

Average surface temperature: 10,000°F (5,500°C)

MERCURY

Tiny Mercury, the planet nearest the Sun, is also the fastest moving. It is named after the speedy, wing-heeled messenger of the Roman gods.

The Extreme Planet

Mercury is the planet with the widest temperature range. During the day, temperatures can reach a sweltering 801°F (427°C), which is twice as hot as an oven. As Mercury has no atmosphere to trap the heat, it escapes back out into space at night. Temperatures plummet to -279°F (-173°C), which is twice as cold as the coldest temperature ever recorded on Earth.

▼ **Mercury's lack of atmosphere prevents meteors from burning up. Instead, they crash into the planet, covering it with impact craters.**

Destination Mercury

From Mercury, the Sun looks three times larger than it does when seen from Earth. However, you would need a very special spacesuit for such a trip. It would not only need to protect you from the vacuum of space and the planet's extreme temperatures, but also from the Sun's deadly radiation, which is stronger on Mercury than on any other planet in the solar system.

Did You Know?

Mercury orbits the Sun at an average speed of 107,082 miles (172,332 km) per hour. This is nearly 188 times faster than a jumbo jet.

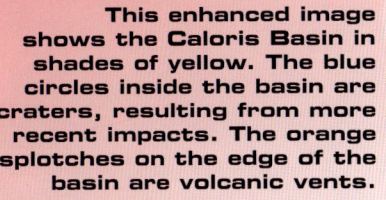

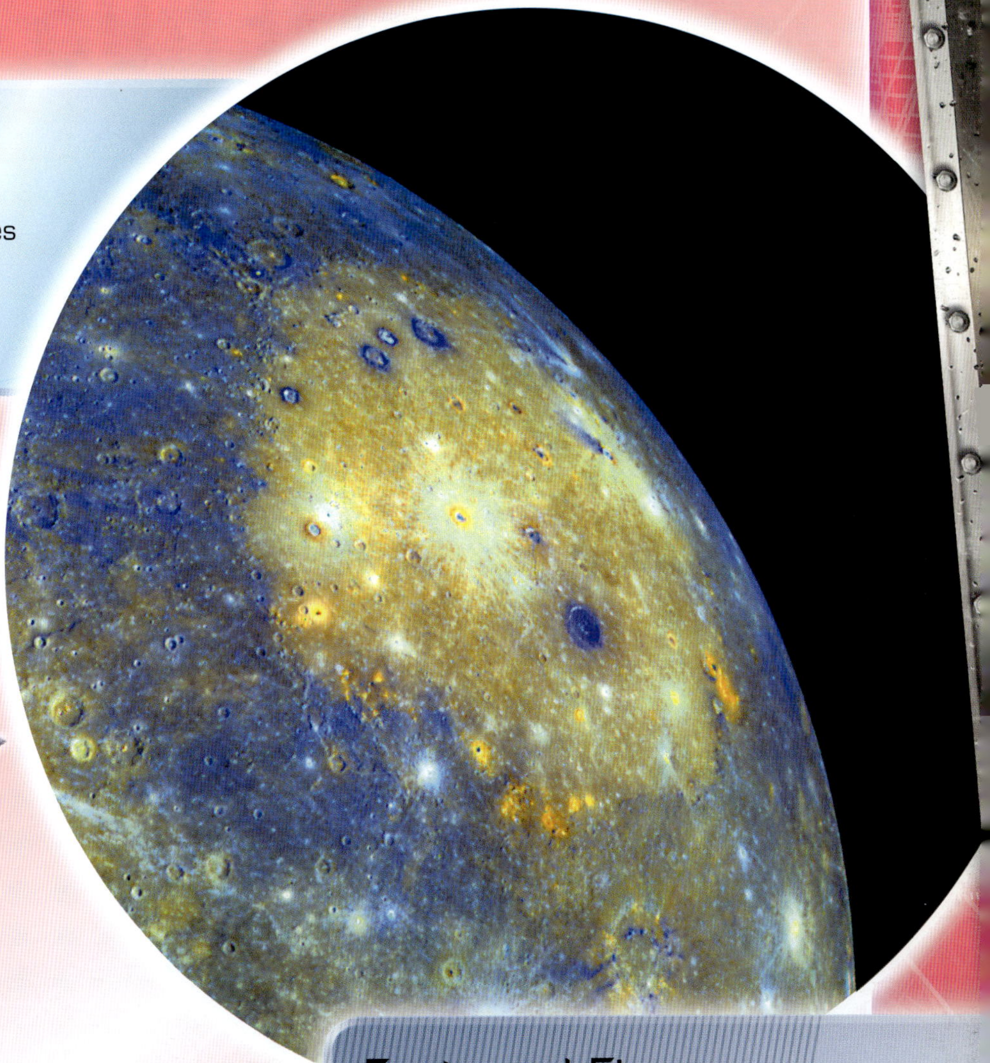

▶ This enhanced image shows the Caloris Basin in shades of yellow. The blue circles inside the basin are craters, resulting from more recent impacts. The orange splotches on the edge of the basin are volcanic vents.

Massive Impact

The largest feature on Mercury is the Caloris Basin, an impact crater 932 miles (1,500 km) across. It was created billions of years ago when a large asteroid crashed into Mercury. On the opposite side of the planet are rocky ridges caused by shock waves from the impact.

Viewing Mercury from Earth

As Mercury is very close to the Sun, it is not easy viewing it from Earth, except at twilight. Mercury also appears indirectly 13 times every century when it passes across the face of the Sun. This event is called a transit. The next transit will occur on May 9, 2016.

Facts and Figures

Name: Mercury

History: known since prehistoric times, named after the messenger of the Roman gods

Type of body: terrestrial planet

Diameter: 3,032 mi (4,879.4 km)

Atmosphere: almost none

Number of moons: 0

Average distance from the Sun: 36 million mi (57.9 million km)

Length of day: 58.646 Earth days

Length of year: 87.97 Earth days

Temperature range: -279–801°F (-173–427°C)

VENUS

Venus is often described as Earth's twin. This is because it is the planet closest to Earth and is also similar in size. In most other ways, the two planets could not be more different.

▲ At sunset, Venus is the largest and brightest planet in the sky.

What is Venus Like?

Venus is an incredibly hostile place. To land there, spacecraft would have to battle hurricane-force winds and pass through thick clouds of sulphuric acid. The surface is as hot as a furnace and the **atmospheric pressure** is strong enough to crush anything that has not already dissolved in the acid or melted in the heat. Venus is the hottest planet in the solar system.

Observing Venus

Venus is the second brightest object in the night sky, after the Moon. It is usually the first shining light in the sky at sunset and the last to disappear at dawn. Venus looks so bright because it is relatively close to Earth, and its thick, white clouds reflect the Sun's light.

Surface Features

Venus is permanently covered in a thick layer of clouds, which hides its surface from view.

The United States's National Aeronautics and Space Administration (NASA) sent the orbiter *Magellan* to Venus in 1989. *Magellan* spent four years **orbiting** the planet and used **radar imaging** to map its surface. It found thousands of volcanoes as well as mountain ranges, canyons, and valleys.

Did You Know?

Scientists use a negative number to express the length of Venus's day. This is because Venus rotates on its axis in the opposite direction from that in which it orbits the Sun. Uranus does the same.

Facts and Figures

Name: Venus

Type of body: terrestrial planet

History: known since prehistoric times, named after the Roman goddess of love and beauty

Diameter: 7,521 mi (12,103.6 km)

Atmosphere: 96.5% carbon dioxide, 3.5% nitrogen

Number of moons: 0

Average distance from the Sun: 67.2 million mi (108.2 million km)

Length of day: -243 Earth days

Length of year: 224.7 Earth days

Average surface temperature: 864°F (462°C)

▼ The photograph on the left shows the thick layer of clouds covering Venus. The radar image on the right reveals the surface of the planet beneath the clouds.

EARTH

Earth is the only place in the solar system where life is known to exist. The planet's atmosphere, oceans, land, and life forms interact to create complex systems scientists are only just beginning to understand.

The Structure of Earth

Earth's outer layer is called the **lithosphere**, which consists of the **crust** and uppermost **mantle** of the planet. Earth's lithosphere is broken up into many **tectonic plates**, which float on the soft, constantly moving rocks of the mantle. Beneath the mantle is Earth's liquid outer core, which is made of iron and nickel. At the center of Earth is the inner core, which is a solid ball of iron and nickel.

▼ This diagram shows the internal structure of Earth. Although the crust is very thin compared to the other layers, even the deepest holes ever drilled have not reached the bottom of the crust.

Facts and Figures

Name: Earth

Type of body: terrestrial planet

Diameter: 7,926.38 mi (12,756.28 km)

Atmosphere: 78% nitrogen, 21% oxygen (and traces of argon, carbon dioxide, and water vapor)

Number of moons: 1

Average distance from the Sun: 93 million mi (149.6 million km)

Length of day: 23.934 hours

Length of year: 365.24 days

Temperature range: -126–136°F (-88–58°C)

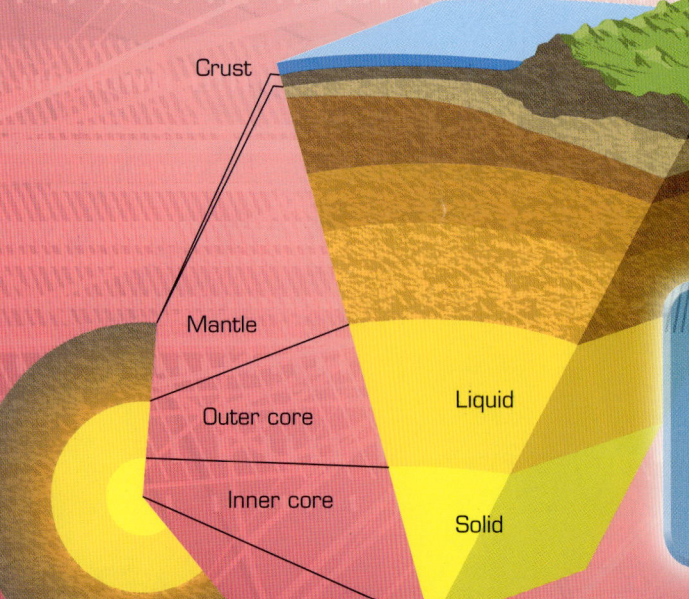

Crust
Mantle
Outer core
Inner core
Liquid
Solid

The lithosphere is the top part of the outer mantle and crust.

Did You Know?

Scientists believe temperatures at the center of Earth's core may be hotter than at the surface of the Sun. They estimate the temperature to be as high as 12,600°F (7,000°C).

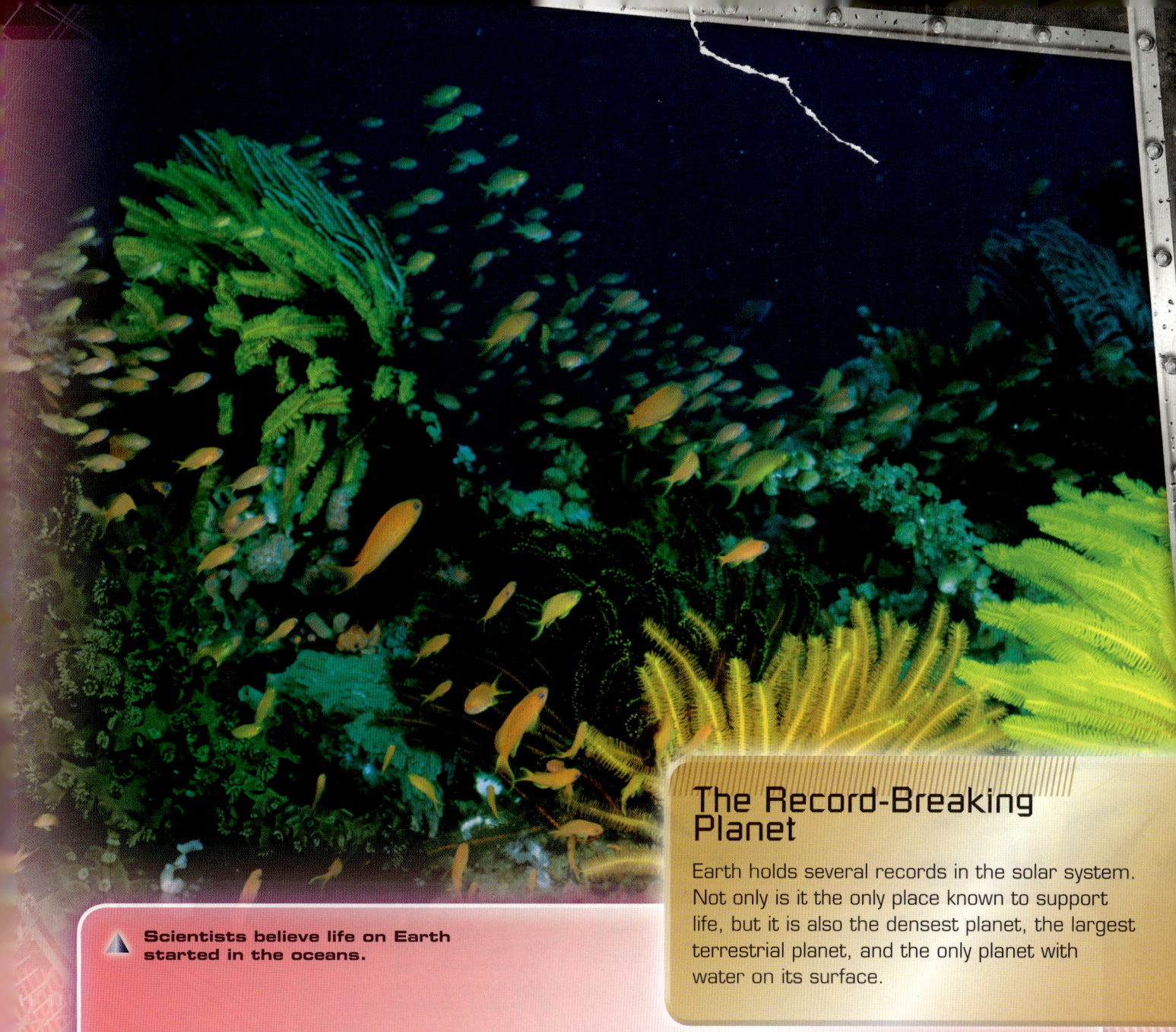

▲ Scientists believe life on Earth started in the oceans.

The Record-Breaking Planet

Earth holds several records in the solar system. Not only is it the only place known to support life, but it is also the densest planet, the largest terrestrial planet, and the only planet with water on its surface.

Why Is There Life on Earth?

Earth is just the right distance from the Sun, so it receives the right amount of sunlight to support life. Earth's atmosphere stops it from overheating during the day and prevents too much heat from escaping at night. This means temperatures on Earth stay within the range where water can remain a liquid. Scientists believe life can only develop in the presence of water.

How are Humans Affecting Earth?

Humans are damaging the planet. They release pollutants into the atmosphere, water, and soil. They consume natural resources more quickly than they can be replaced and destroy natural habitats. Humans have caused or contributed to a variety of problems, including acid rain, the depletion of the ozone layer, global warming, and the extinction of many species of plants and animals.

THE MOON

The Moon is Earth's closest neighbor in space and its only natural satellite. The Moon is lit by reflected sunlight and is the brightest object in the night sky.

Observing the Moon

The dark regions of the Moon are called *mare*, which is Latin for sea. They are actually impact basins formed when objects crashed into the Moon. Dark lava eventually filled these basins. The lighter regions are called *terra*, meaning land. They are rugged mountain ranges. The Moon is covered in craters, many of them more than 25 miles (40 km) wide.

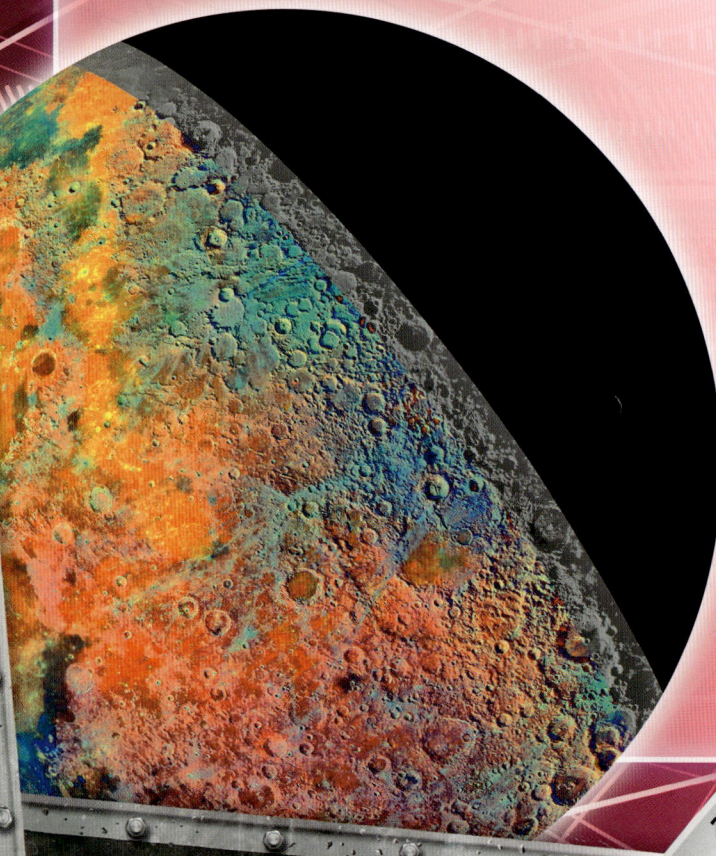

▼ This false color image of the Moon shows its surface features. Dark blue to orange shades indicate ancient lava flows, which filled the *mare*. Bright pink to red shades represent the highlands, while the youngest craters have light blue rays extending from them.

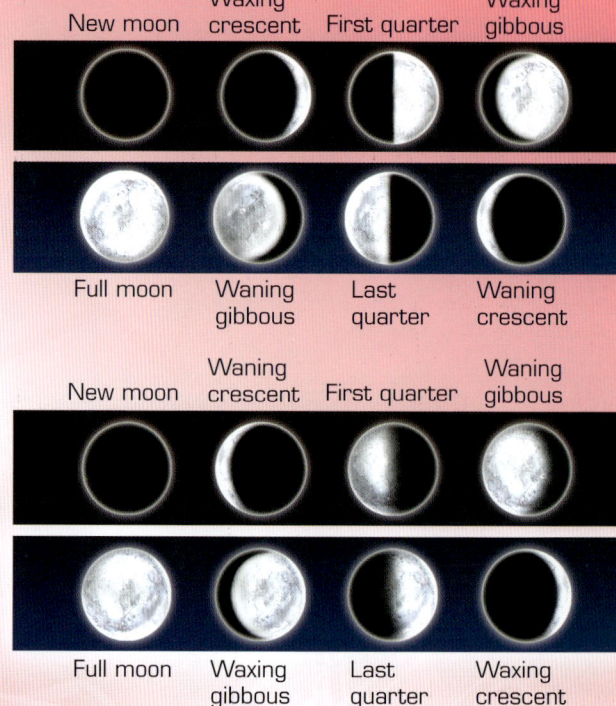

▲ As the Moon orbits Earth, it appears to change shape because different parts of its sunlit half face Earth. The shapes above show what the Moon looks like from Earth in the Northern (top) and Southern (below) hemispheres.

How Does the Moon Affect Earth?

Even though the Moon is very small and distant, it still affects Earth. The Moon's gravity steadies the planet and its climate. Earth wobbles on its axis and, as it does so, its angle toward the Sun changes. This alters the amount of sunlight Earth receives, which affects its climate. The Moon's gravity pulling on Earth steadies the wobble and stabilizes the climate, making conditions more favorable for life. The Moon's gravity also pulls the water in the oceans, causing the tides.

Apollo 17 astronaut Harrison Schmitt is pictured here on the moon with the lunar rover. This color photograph also shows the Moon's black sky and grey surface.

Did You Know?

The Moon rotates at roughly the same speed as it orbits Earth. As a result of this, the same side of the Moon always faces Earth. No one knew what the far side of the Moon looked like until the **Soviet Union**'s *Luna 3* space probe took photos of it in 1959.

What is the Moon Like?

Although the Moon shares Earth's orbit around the Sun, it is a very different world to Earth. Unlike Earth, the Moon has no protective atmosphere. As a result, meteors do not burn up before impacting the Moon's surface. Instead, they crash into it, leaving the surface heavily cratered. The lack of atmosphere also means temperatures are more extreme on the Moon than on Earth. The average daytime temperature is 253°F (123°C), plummeting to around -387°F (-233°C) at night.

Facts and Figures

Name: the Moon

Type of body: planetary satellite

History: known since prehistoric times

Diameter: 2,159.1 mi (3,474.8 km)

Atmosphere: none

Average distance from Earth: 238,855 mi (384,400 km)

Length of day: 27.3 Earth days

Length of year: 365.24 Earth days

Temperature range: -387–253°F (-233–123°C)

17

MARS

Mars is the fourth planet from the Sun and the outermost of the terrestrial planets. It is known as the Red Planet, because of its red-orange color. The Martian soil is red because it contains iron oxide, or rust.

Mars Then and Now

Mars is a cold, dry planet. There is no water on its surface, but there is plenty trapped just below the surface in the form of ice. Mars has not always been cold and dry. Several billion years ago, it was much warmer and had rivers, lakes, and seas. Scientists are still trying to find out what happened to make Mars change so dramatically.

Features of Mars

Mars is home to the largest volcano in the solar system—Olympus Mons. It is 372.8 miles (600 km) wide and 16.8 miles (27 km) high, which is more than three times the height of Mount Everest, the highest mountain on Earth.

Mars also has the longest known canyon in the solar system—Valles Marineris. It is more than 2,500 miles (4,000 km) long, which makes it almost nine times the length of the Grand Canyon in the United States, the longest canyon on Earth.

▼ This photograph was taken on Mars by NASA's *Phoenix Mars Lander*. It shows part of the spacecraft's solar panels and its robotic arm, with a sample of Martian soil in the scoop.

Did You Know?

Mars has two moons called Phobos and Deimos, which scientists believe are asteroids that were trapped by the planet's gravity. They are much smaller than Earth's Moon. Phobos is 13.8 mi (22.2 km) wide and Deimos is just 7.8 mi (12.6 km) wide.

Is There Life on Mars?

For a long time, it was believed that there might be intelligent life on Mars. Nowadays, scientists believe early conditions on the planet may have allowed the evolution of simple life forms, such as bacteria, but so far they have found no evidence to support this theory. Spacecraft currently exploring Mars are still looking for signs of life.

Facts and Figures

Name: Mars

Type of body: terrestrial planet

History: known since prehistoric times, named after the Roman god of war

Diameter: 4,333.4 mi (6,974 km)

Atmosphere: 95.3% carbon dioxide, 2.7% nitrogen, 1.6% argon (and traces of carbon monoxide and water vapor)

Number of moons: 2

Average distance from the Sun: 141.6 million mi (227.9 million km)

Length of day: 1.026 Earth days

Length of year: 686.93 Earth days

Temperature range: -125–23°F (-87– -5°C)

▼ The *Hubble Space Telescope* captured this series of images of Mars. Each view shows a different region of Mars as the planet completes one-quarter of its daily rotation.

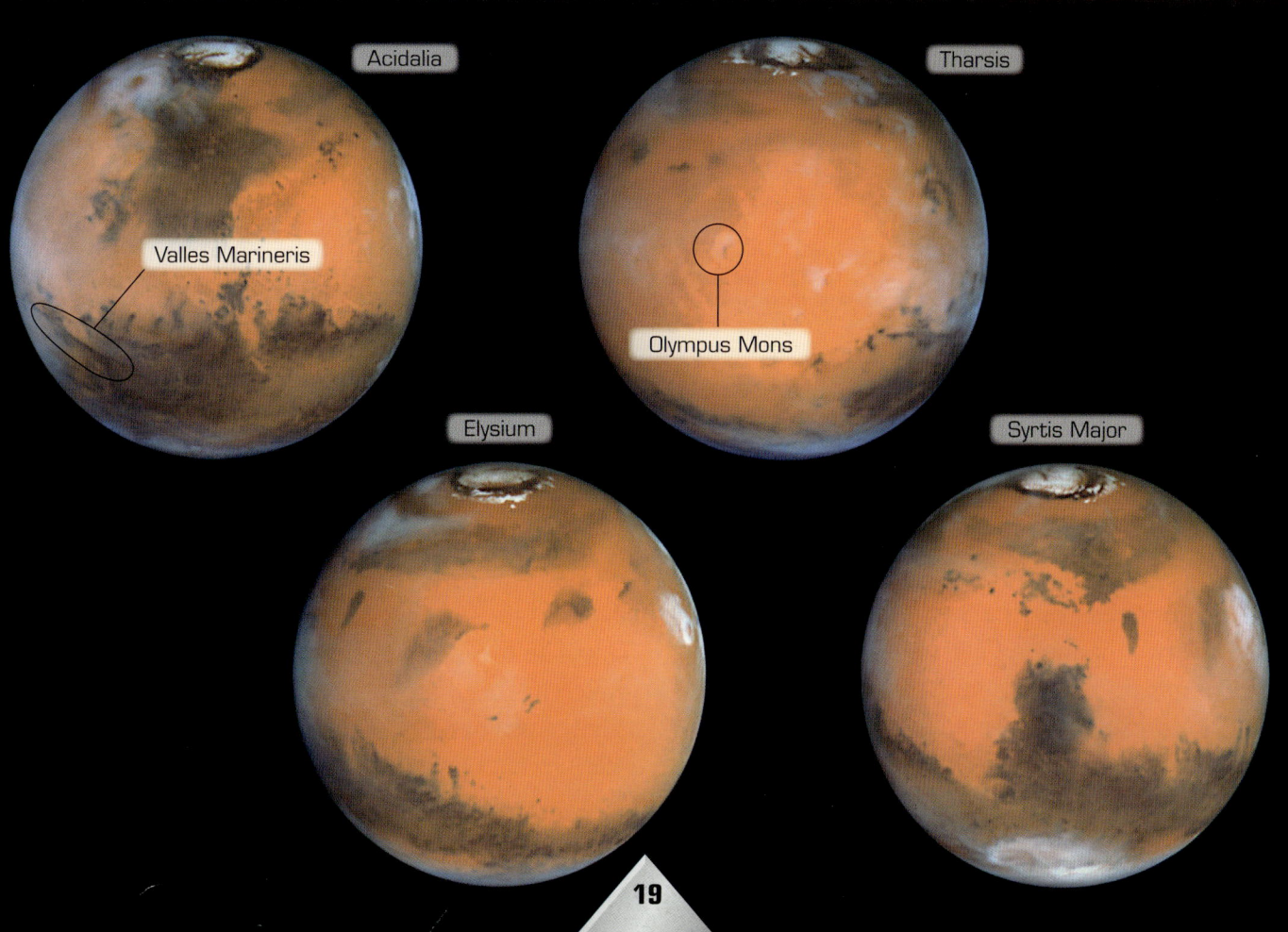

JUPITER

Jupiter is by far the largest planet in the solar system. It is 2.5 times as massive as all the other planets combined.

▲ This close-up view of Jupiter was taken by NASA's *Voyager 1* spacecraft. The large orange blob near the top of the image is the Great Red Spot.

The Jupiter System

Jupiter and its moons could be described as a mini solar system. Jupiter has four large moons, which are the size of planets, and numerous smaller ones. Like the Sun, Jupiter has a powerful magnetic field. Its **magnetosphere** extends up to 1.8 million miles (3 million km) toward the Sun and more than 621.4 million miles (1 billion km) away from it.

Features of Jupiter

Jupiter is a gas giant rather than a terrestrial planet. This means it is mainly made of different gases rather than solid matter. The swirly patterns seen in Jupiter's upper atmosphere are clouds of water and a toxic gas called ammonia. The most prominent feature on the planet is the Great Red Spot. It is a giant spinning storm up to three times the diameter of Earth, which has been raging for at least 300 years. Jupiter also has a faint ring system, discovered in 1979 by NASA's *Voyager 1* spacecraft.

The Galilean Satellites

Jupiter's four largest moons are called the Galilean satellites, named after the Italian astronomer who discovered them, Galileo Galilei.
- Ganymede is the largest moon in the solar system (larger than the planet Mercury).
- Io is the most volcanically active moon in the solar system.
- Europa may have liquid water beneath its icy crust.
- Callisto's ancient surface could reveal a lot about the early solar system.

Did you know?

Jupiter contains roughly the same mix of gases as the Sun. If Jupiter had about 80 times more mass, it would be a star instead of a planet.

▼ These are the Galilean satellites. The relative sizes of the moons are shown to scale, but the distances between them are not.

Facts and Figures

Name: Jupiter

Type of body: gas giant

History: known since prehistoric times, named after the king of the Roman gods

Diameter: 88,846 mi (142,984 km)

Volume: 1,316 Earths could fit inside

Composition: 90% hydrogen, 10% helium (and traces of methane, ammonia, water, and other gases)

Number of moons: 63

Number of rings: 3

Average distance from the Sun: 483.7 million mi (778.4 million km)

Length of day: 0.41354 Earth days

Length of year: 4,330.6 Earth days

Average temperature: -234°F (-148°C)

Io   Europa   Ganymede   Callisto

Saturn

The gas giant Saturn, the second largest planet in the solar system, has been known since prehistoric times. Its rings, however, were only discovered in 1610, when Italian astronomer Galileo Galilei looked at the planet, using a new invention called a telescope.

What is Saturn Like?

Saturn is a lot like Jupiter. It is made of a similar mix of gases, has super-strong winds and its atmospheric pressure is very high. Saturn is also the only planet in the solar system that is less dense than water. This means it would float if placed in water.

Saturn's Rings

Saturn has a spectacular ring system. This system is made up of billions of particles of ice and rock. Scientists are not sure how the rings formed. A possible explanation is that one of Saturn's icy moons, or a comet or asteroid, came too close to the planet and was torn apart by its powerful gravity.

▼ The *Cassini* orbiter took this spectacular image of Saturn and its rings in 2008.

Did You Know?

The pieces of ice and rock that make up Saturn's rings vary greatly in size. Some are as small as a grain of sugar, and others are as big as a house.

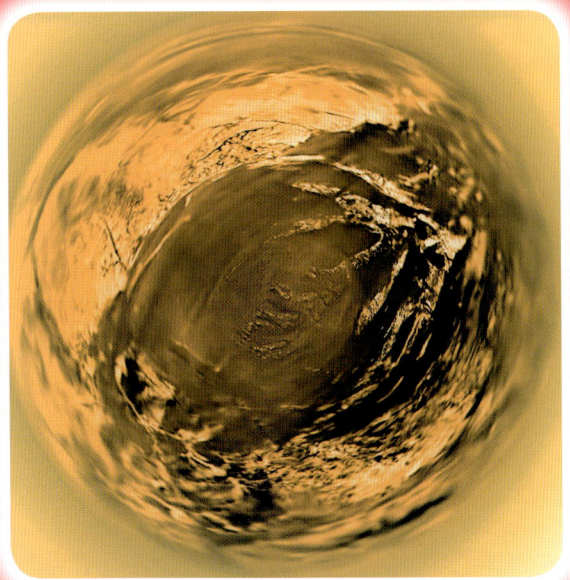

▲ This view of Titan's surface was captured by the *Huygens* space probe from a height of 3.1 mi (5 km). It was taken with a fish-eye lens.

Facts and Figures

Name: Saturn

Type of body: gas giant

History: known since prehistoric times, named after the Roman god of agriculture

Diameter: 74,898 mi (120,536 km)

Volume: 763 Earths would fit inside

Composition: 96% hydrogen, 3% helium (and traces of methane, ammonia, and other gases)

Number of moons: 61

Number of rings: thousands

Average distance from the Sun: 886 million mi (1.4 billion km)

Length of day: 0.44401 Earth days

Length of year: 10,755.7 Earth days

Average temperature: -288°F (-178°C)

Saturn's Moons

So far, 61 moons have been discovered in orbit around Saturn, ranging in diameter from 1.2 miles (2 km) to more than 3,107 miles (5,000 km). The table below describes four of the largest moons.

Name	Diameter	Features
Titan	3,200 mi (5,150 km)	It is the only moon known to have a planet-like atmosphere and liquid on its surface. Titan's rivers, lakes, and seas consist of liquid methane, the main component of natural gas.
Iapetus	914 mi (1,471 km)	• It is shaped like a walnut. • Half of the moon is very dark and the other half is very bright.
Enceladus	307 mi (494 km)	It has a **hot spot**, which sends enormous plumes of ice and water gushing into space. It is believed that the material ejected from this hot spot created Saturn's Ring E.
Mimas	244 mi (392 km)	The largest impact crater on Mimas is almost one-third the diameter of the moon itself. Mimas is responsible for clearing the ring material from the Cassini Division, the gap between Saturn's two widest rings. Mimas's gravity pulled all the material into it.

Uranus

Uranus was the first planet to be discovered in modern times. Although ancient astronomers observed it, they did not realize it was a planet, because it was so faint and moved so slowly.

How was Uranus Discovered?

In 1781, British–German astronomer William Herschel discovered what he thought was a comet. When other astronomers studied this comet, they found that it was, in fact, a planet. Herschel named the new planet Georgium Sidus (George's Star) after King George III of the United Kingdom. However, the name was not popular and was eventually changed to Uranus.

Did You Know?

Uranus is tilted on its side. This means its north and south poles lie where other planets have their equators. Rather than rotating like a spinning top, Uranus rotates like a rolling ball. This situation may have been caused by a collision with another planet long ago.

▼ This diagram shows the tilt of Uranus and how its rings have appeared and will appear between 1965 and 2028.

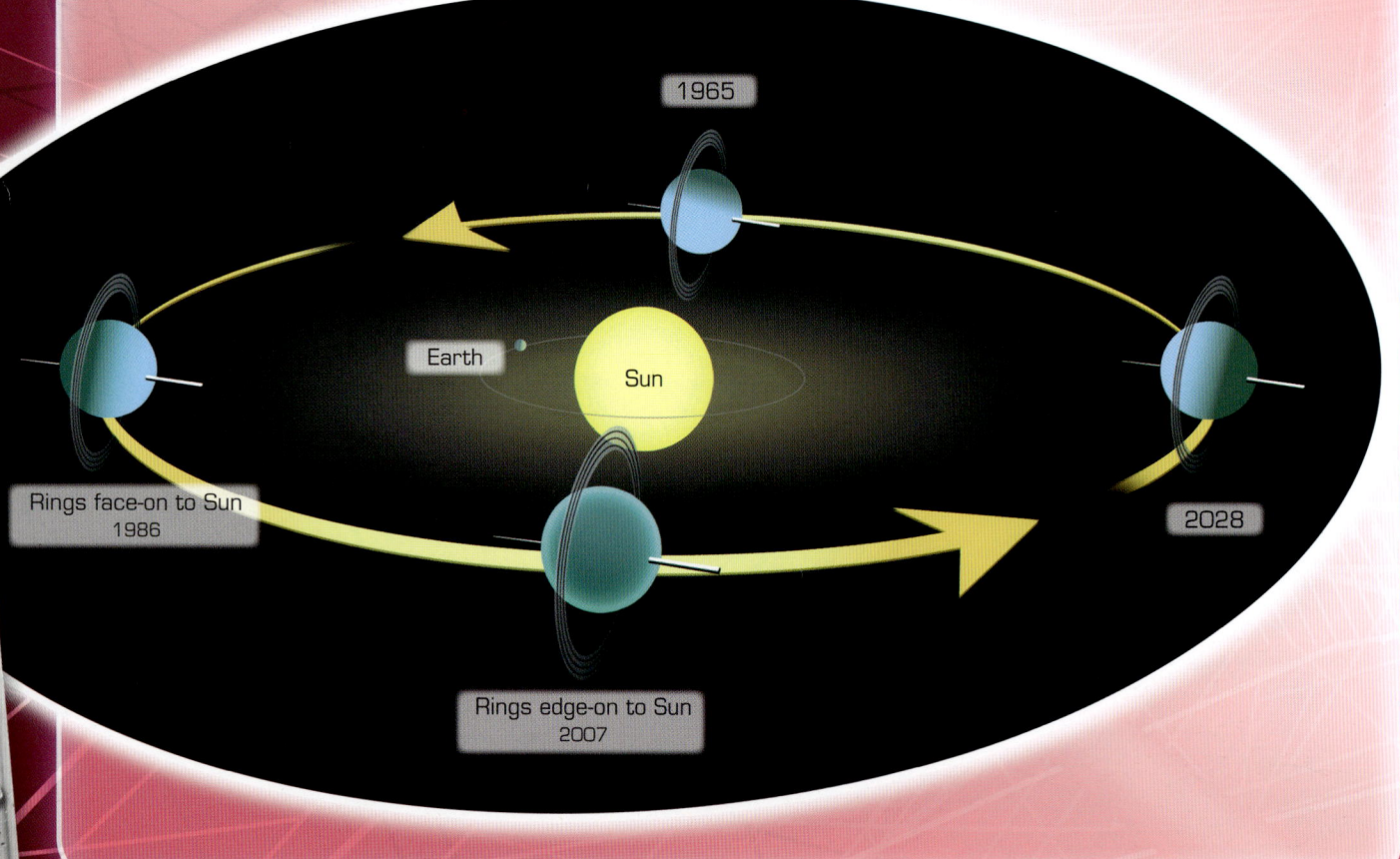

This combined image of Uranus and its five major moons consists of images taken by the *Voyager 2* spacecraft.

What is Uranus Like?

Uranus is a gas giant with a complex, layered cloud structure. The upper clouds are mainly composed of methane, which gives the planet its blue color. The lower clouds are mainly made of water. Uranus might seem calm and peaceful due to its smooth, consistent appearance, but appearances can be deceptive. Wind speeds on Uranus can reach up to 600 miles (900 km) per hour.

Uranus's Moons and Rings

Uranus has 27 known moons. The most intriguing of these is Miranda, which has a rather messy appearance. Some scientists believe Miranda may have been shattered a number of times during its life span, with its broken parts eventually coming back together, resulting in its current form.

Uranus also has a ring system, which was only discovered in 1977. Its rings are mostly dark and narrow, and no more than 33 feet (10 m) thick. It is hardly surprising they were difficult to spot from Earth, 1.8 billion miles away!

Facts and Figures

Name: Uranus

Type of body: gas giant (also classified as an ice giant)

History: discovered in 1781 by William Herschel, named after the Greek god of the heavens

Diameter: 31,763 mi (51,118 km)

Volume: 63 Earths would fit inside

Composition: 82.5% hydrogen, 15.2% helium, and 2.3% methane

Number of moons: 27

Number of rings: 13

Average distance from the Sun: 1.8 billion mi (2.9 billion km)

Length of day: -0.7916 Earth days

Length of year: 30,687.2 Earth days

Average temperature: -357°F (-216°C)

NEPTUNE

Neptune, the solar system's outermost planet, was the first planet to be discovered through mathematical calculations rather than simple observations of the night sky.

How Was Neptune Discovered?

Galileo Galilei classified Neptune as a star between 1612 and 1613, but it was not until 1846 that Neptune was officially discovered and classified as a planet.

It was once believed that Uranus was the planet farthest from Earth. However, when astronomers noticed that Uranus was not orbiting the Sun as it was expected to, they theorized that there must be another planet affecting it from an orbit farther out.

French mathematician Urbain Le Verrier calculated the position of this missing planet in mid-1846 and sent the information to German astronomer Johann Galle. Galle found the new planet on the first night he looked for it.

What Is Neptune Like?

Neptune, another gas giant, is the windiest planet in the solar system, with wind speeds of more than 1,243 miles (2,000 km) per hour. NASA's *Voyager 2* spacecraft observed one particularly large storm in 1989, which scientists named the Great Dark Spot. It was as big as Earth and moved across Neptune at speeds of more than 684 miles (1,100 km) per hour.

Voyager 2 captured this image of three storms on Neptune.

Great Dark Spot

Scooter

Dark Spot 2

Did You Know?

Neptune's winds are three times stronger than Jupiter's and nine times stronger than winds on Earth.

▲ Triton is the coldest object in the solar system, with a surface temperature of -391°F (-235°C). Its surface is made mainly of nitrogen ice. The pinkish areas may contain methane ice. The dark streaks are probably deposits from geysers.

Neptune's Moons and Rings

There are 13 moons orbiting Neptune. The largest moon, Triton, ejects huge plumes of liquid, which instantly freeze and fall back down as snow. Triton is the only moon that orbits its planet in a direction opposite to the planet's rotation. This suggests that Triton may have been a dwarf planet that was captured by Neptune's gravity.

Neptune also has five faint rings, which are believed to be disintegrating.

Facts and Figures

Name: Neptune

Type of body: gas giant (also classified as an ice giant)

History: discovered in 1846 by Urbain Le Verrier and Johann Galle, named after the Roman god of the sea

Diameter: 30,775 mi (49,528 km)

Volume: 57.7 Earths would fit inside

Composition: 80% hydrogen, 19% helium, 1.% methane

Number of moons: 13

Number of rings: 5

Average distance from the Sun: 2.8 billion mi (4.5 billion km)

Length of day: 0.67126 Earth days

Length of year: 60,190 Earth days

Average temperature: -214°C (-353°F)

Other Bodies

Planets and their moons are not the only objects in orbit around the Sun. Our solar system is made up of millions of smaller bodies, too, including asteroids, dwarf planets, and comets.

Asteroids

Asteroids are fragments of rock which remained after the formation of the solar system. Most asteroids can be found in the **asteroid belt**, between Mars and Jupiter. The largest asteroids are up to one-quarter the size of the Moon. The smallest asteroids are less than 0.6 miles (1 km) across. There are more than 90,000 asteroids in the solar system.

Dwarf Planets

Like a planet, a dwarf planet is roughly round in shape and orbits the Sun directly. However, a dwarf planet is not large enough to have cleared its orbital space of smaller bodies. Five dwarf planets have been discovered so far. Ceres lies in the asteroid belt. Pluto, Haumea, Makemake, and Eris are beyond Neptune, in a region called the Kuiper Belt.

▼ **This artist's impression shows the dwarf planet Pluto with its moon, Charon.**

Did You Know?

Pluto was originally classified as a planet. When astronomers discovered several more distant objects in the solar system that were larger than Pluto, they decided to revise their definition of a planet. In 2006, Pluto was reclassified as a dwarf planet.

Pluto

Charon

28

▲ Comet Hyakutake was discovered in 1996 and is believed to have the longest tail known for a comet.

Comets

Just like asteroids, comets are pieces of debris which remained after the formation of the solar system. However, unlike asteroids, comets contain a large proportion of ice, and most of them have extended, off-center orbits. Comets travel to the outer reaches of the solar system at one end of their orbit and very close to the Sun at the other end.

When a comet gets close to the Sun, the Sun's heat causes ice on the comet to turn into gas, causing the comet to glow. The solar wind blows the glowing gas away from the Sun, creating the comet's distinctive tail.

Did You Know?

When a meteoroid enters Earth's atmosphere, it is called a meteor. Any part of a meteor that hits the ground is called a meteorite.

Other Space Bodies

Meteoroids are rocks hurtling through space. They may be asteroids, comets, pieces of comets, or chunks knocked off other planets during a collision. Between 2.2 and 22 million pounds (1–10 million kg) of meteoroid material fall to Earth every day, mostly as dust-sized particles. Scientists believe many **mass extinctions** may have been caused by much larger impacts, including the event that wiped out the dinosaurs.

The Future of Our Solar System

Our solar system formed from a solar nebula, a disc of dust and gas left behind when the Sun was formed. The solar system has changed a lot in the first 4.6 billion years of its existence. How is it likely to change in the future?

Getting Hotter

The future of our solar system is closely linked to the fate of the Sun. Over the next 5 billion years or so, the Sun will gradually become hotter and hotter. In about 500 million years, Earth will too hot to support life. In about 1 billion years, the surface of the planet will be so hot that all water will evaporate.

A New Beginning?

Just as life on Earth follows the circle of life, the solar system does the same. When conditions on Earth are no longer able to support life, it is possible that changing conditions on planets or moons farther away from the Sun, such as Mars or the moons of Jupiter, may allow life to evolve all over again.

▼ The planets, stars, and moons in our solar system are evolving, aging, and may one day end. However, new ones are forming elsewhere, as seen in this artist's impression of an Earth-like planet forming around a star.

GLOSSARY

asteroid belt
a region between the orbits of Mars and Jupiter where most of the asteroids in the solar system can be found

asteroids
rocky or metallic objects that vary in size from a few feet to more than 560 miles (900 km) across and orbit the Sun or another star

atmosphere
the layer of gases surrounding a planet, moon, or star

atmospheric pressure
the force at any given point in a planet's atmosphere, created by the weight of the atmosphere's gases above it

comets
collections of ice, dust, and small, rocky particles that orbit the Sun

crust
the outer layer of a rocky planet or moon

dwarf planets
small planet-like bodies that are not satellites of another body but still share their orbital space with other bodies

electromagnetic radiation
waves of energy created by electric and magnetic fields

elements
pure chemical substances which cannot be reduced to a simpler form

galaxy
a large system of stars, gas, and dust held together by gravity

gas giants
large planets made mostly of gas and with a metal or rock core, such as Jupiter, Saturn, Uranus, and Neptune

gravity
the strong force that pulls one object toward another

hot spot
an area of active volcanism on the surface of a planet or moon

impact craters
craters formed on a planet or moon when smaller bodies crash into it

lithosphere
the solid outer shell of a rocky planet or moon

magnetosphere
the region around a planet or star that is affected by its magnetic field

mantle
the layer of a rocky planet or moon immediately below the crust

mass extinctions
a sharp fall in the number of species in a short period of time

meteors
pieces of space rock that get caught by the gravity of a moon or planet and fall toward it

moons
natural bodies which orbit planets or other bodies

nuclear fusion
the process in which the nuclei of two or more atoms fuse together to form a single atom with a heavier nucleus, releasing huge amounts of energy

orbited
followed a curved path around a more massive object while held in place by gravity; the path taken by the orbiting object is its orbit

radar imaging
a technique that creates images using radio waves instead of visible light

radiation
energy emitted in the form of waves, such as electromagnetic radiation, which may be harmful to living things

red giant
a very large, cooling star

satellite
a natural or artificial object in orbit around another body

solar system
the Sun and everything in orbit around it, including the planets

Soviet Union
a nation that existed from 1922 to 1991, made up of Russia and 14 neighboring states

tectonic plates
large, thin, and relatively rigid plates that move against each other and make up the upper layer of Earth's crust

white dwarf
a small, dense star that is very faint and in the final stage of its evolution

INDEX

A
asteroid belt 28, 31
asteroids 5, 11, 19, 22, 28, 31
atmospheric pressure 12, 22, 31

C
Callisto 21
Caloris Basin 11
canyons 13, 18
Ceres 28
comets 5, 22, 24, 28, 29, 31
core 8, 9, 14
coronal mass ejections (CMEs) 9
craters 10, 11, 16, 17, 23

D
dwarf planets 5, 27, 28–29, 31

E
Earth 4, 5, 6, 8, 10, 12, 14–15, 16, 17, 18, 19, 20, 25, 26, 29, 30
Earth's wobble 16
electromagnetic radiation 8, 31
elements 5, 31
Enceladus 23
Eris 28
Europa 21

F
far side of the Moon 17

G
Galilean satellites 21
Galilei, Galileo 21, 22, 26
Galle, Johann 26, 27
Ganymede 21
gas giant 5, 20, 22, 25, 26, 31
global warming 15
gravity 5, 16, 19, 22, 23, 27, 31
Great Dark Spot 26
Great Red Spot 20

H
Haumea 28
Herschel, William 24, 25
hot spot 23, 31

I
Iapetus 23
impact craters 10, 11, 23, 31
Io 21

J
Jupiter 5, 6, 20–21, 22, 26, 28, 30

K
Kuiper belt 28

L
Le Verrier, Urbain 26, 27
lithosphere 14, 31
Luna 3 17

M
Magellan 13
magnetosphere 20, 31
main sequence 8
Makemake 28
mantle 14, 31
mare 16
Mars 5, 6, 13, 18–19, 28, 30
Mercury 5, 6, 10, 11
meteorites 29
meteoroids 29
meteors 17, 29, 31
Milky Way 5, 8
Mimas 23
Miranda 25
moons 5, 19, 20, 21, 22, 23, 25, 27, 28, 30, 31
Moon, the 4, 12, 16–17, 28
mountain ranges 13, 16

N
National Aeronautics and Space Administration (NASA) 13, 18, 20, 26
Neptune 5, 6, 26–27, 28
nuclear fusion 8, 31

O
Olympus Mons 18

P
plasma 9
Pluto 28

R
radar imaging 13, 31
radiation 8, 10
red giant 8, 31
rings 22, 23, 25, 27

S
satellite 16, 21, 31
Saturn 5, 6, 22–23
Sedna 5
solar nebula 5, 30
solar system 4, 5, 6–7, 10, 12, 14, 15, 18, 20, 21, 22, 26, 27, 28, 29, 30, 31
solar wind 29
Soviet Union 17, 31
stars 5, 8, 21, 26, 30
Sun, the 5, 6, 8–9, 10, 11, 12, 13, 14, 15, 16, 17, 18, 20, 21, 26, 28, 29, 30

T
tectonic plates 14, 31
terra 16
Titan 23
Triton 27

U
Uranus 5, 6, 13, 24–25, 26

V
Valles Marineris 18
valleys 13
Venus 5, 6, 12–13
volcanoes 13, 18
Voyager spacecraft 20, 26

W
white dwarf 8, 31